2/96

D0573749

# HURRICANES

CHARLES ROTTER

CREATIVE EDUCATION

Designed by Rita Marshall
with the help of Thomas Lawton

Published by Creative Education
123 South Broad Street,
Mankato, Minnesota 56001
Creative Education is an imprint
of Creative Education, Inc.

Photography by Peter Arnold, Inc.
(NASA), Tony Arruza, Comstock
(Jack K. Clark, Cameron Davidson,
Franklin Viola), FPG International
(Jeffrey Sylvester), DRK Photo
(Annie Griffiths), H. Armstrong Roberts
(E.R. Degginger, K. Scholz),
International Stock (Warren Faidley),
Tom Stack & Associates (Scott
Blackman, Jack Stein Grove, Brian
Parker, F. Stuart Westmorland), and
Tony Stone Worldwide (Ken Biggs)

Library of Congress
Cataloging-in-Publication Data

Rotter, Charles.
Hurricanes / Charles Murray
Rotter.
Summary: Discusses the definition,
formation, destructive power, and
forecasting of hurricanes or tropical
cyclones.
ISBN 0-88682-597-0
1. Hurricanes—Juvenile
literature.   2. Hurricane Camille,
1969—Juvenile literature.
3. Cyclones—Tropics—Juvenile
literature.   [1. Hurricanes.]
I. Title.                       92-44442
QC994.2.R68   1993        CIP
551.55' 2—dc20              AC

In Memory of
GEORGE R. PETERSON, SR.

## 6

*Hurricanes represent a terrifying side of* nature. Striking various areas around the globe, they have been wreaking havoc longer than people have been walking the earth. These immense storms batter the landscape with their fierce winds and torrential rains. The awful specter of destruction can haunt people for a long time—far longer than it takes to rebuild the actual damage done by the waves, wind, and rain.

*The menacing waves of a hurricane.*

Hurricanes have different names in different parts of the world. Only when they originate in the Atlantic Ocean or the eastern Pacific Ocean do we call them hurricanes. On the western Pacific, off the coast of China, for example, they are called *Typhoons,* from a Chinese phrase meaning "big wind." When they originate in the Indian Ocean, they are called *Cyclones.* Australians probably have the most colorful name for hurricanes: They call the storms *Willy-Willies.*

Every hurricane, no matter what its name or where it occurs, is a type of weather disturbance known as a *Tropical Cyclone.* Usually about 300 or 400 miles (483 or 644 km) in diameter, the storms can generate winds exceeding 200 miles (322 km) per hour. This wind speed is part of the definition of a hurricane. If the wind speed is less than 75 miles (121 km) per hour, the storm does not qualify as a hurricane.

*The winter surf along the coast of Oregon.*

Tropical cyclones are caused by changes in the *Atmosphere,* the layer of air that surrounds the earth. Photographed from space, the earth seems covered with still blue oceans and wispy white clouds. But this tranquil appearance is deceiving. The earth's atmosphere is very active, constantly moving and changing both its temperatures and its moisture patterns. These changes cause everything we call weather, including rain, snow, thunderstorms, tornadoes, and, of course, hurricanes. Even the most violent winds are just rapidly moving air.

*Earth as seen from space.*

## 12

The energy source that drives this air movement is the sun. As the earth orbits the sun, a great deal of light and heat shine down upon the planet. Some of this energy bounces back into space, but most of it is absorbed by the lands and oceans of the earth's surface. The greatest warming occurs near the equator, where the sun normally shines the brightest.

*Atmospheric changes over the ocean.*

## 13

As the sun heats the earth's surface, it also heats the air in contact with that surface. The heated air expands and becomes lighter. Lighter air rises, just like the heated air that lifts a hot air balloon. This creates an area called a *Low-Pressure Zone*, where the air is less dense than the air that surrounds it. Such zones are also called *Depressions*. When a mass of air is cool and dense, it forms a *High-Pressure Zone*.

A difference in air pressure causes air to flow from the higher pressure zone to the lower one to equalize the difference. This difference is known as the *Pressure-Gradient*, and the moving air is what we call wind. The wind does not move in a straight path. Instead, it flows at an angle to the pressure-gradient. This is due to the *Coriolis Effect*, a phenomenon discovered in 1835 by French scientist Gaspard-Gustav de Coriolis. By mathematical calculations Coriolis showed that a moving object's path on a rotating surface curves away from its original direction. This explains why a ball rolled across the surface of a spinning merry-go-round will move in a curve.

*A storm system moves across the Sacramento Valley, California.*

Similarly, air moving inward toward a depression moves in a circular path. This spiral of air is called a *Cyclone,* a term invented by Captain Henry Piddington in the late 1830s. Piddington was studying storms off the coast of India when he noticed a whirling pattern in the clouds. He called the storms cyclones, from the Greek word *kyklon,* meaning "whirling around." Modern technology allows us to appreciate how good his choice of name was: Now, when we use radar to map hurricanes or when we view them from space, we can easily see the whirling pattern.

Today we use the term cyclone to describe not just the storms in the Indian Ocean, but any mass of air spiraling in toward a depression. In the Northern Hemisphere, cyclones turn counterclockwise; in the Southern Hemisphere, they turn clockwise.

*Hurricane Gladys as seen from Apollo 7, 1968.*

A hurricane is a special type of cyclone called a tropical cyclone. Tropical cyclones form over warm seas, far from land. The ocean surface must be warm enough to provide the energy a cyclone needs—at least 80 degrees Fahrenheit (27 C). At these temperatures, warm, moist air constantly rises from the surface of the water to create low-pressure regions over the open ocean. Cool, denser air spirals into the depression, creating the cyclone wind pattern. Only about one in ten of these tropical depressions will attain the energy needed to become a full-fledged tropical cyclone, or hurricane.

*A storm front over the Great Barrier Reef, Australia.*

Appearing both above and below the equator, tropical cyclones originate between 5 and 20 degrees latitude. Within 5 degrees of the equator, the Coriolis effect is too small to provide the spin needed to build large storms; outside of 20 degrees, the water isn't warm enough. After forming, the storms tend to move in a westward direction. This means that land at the western edges of oceans is much more threatened by hurricanes than land at the eastern edges.

*Rough waters on the open ocean.*

The strong winds of a hurricane spiral inward toward the center of the storm, called the *Eye*. Ranging from 5 to 25 miles (8 to 40 km) wide, the eye of a hurricane is calm and almost windless. This calm can create an extremely hazardous situation: Thinking that the storm has passed, people may relax and leave shelters. But the danger is far from over—it is actually only moments away from its peak.

*Dawn at sea.*

Hurricanes can cause incredible amounts of damage when they move onto land. Storm-swollen waves do the most harm. The low-pressure center at the eye of a hurricane can cause the sea level to rise several feet. Driven by the violent wind, the waves batter the shore and flood low-lying communities near the coast.

*Wind-driven waves.*

As the warm, moist air of a hurricane rises and condenses, the storm also generates intense rainfall, which makes the flooding even worse as the storm passes onto land. When moisture condenses into water, heat energy is released. The massive rainfall from a hurricane releases a great deal of heat energy, which continues to drive the storm on land even though it has been deprived of its main source of energy (the rising moist air of the tropical seas). But this new source of power is limited, and while over land, the hurricane steadily weakens. It can regain its strength only if it moves out over the open water. Otherwise, it will continue to weaken until it is reduced to an ordinary tropical depression.

*Hurricanes can affect inland weather patterns.*

In 1969, one of the mightiest hurricanes in the modern history of the Western Hemisphere lashed the Atlantic coast of the United States. First spotted by a weather satellite on August 5, *Hurricane Camille* formed near the Cape Verde Islands, off the northwest coast of Africa. The storm grew in size until it reached full hurricane proportions, then headed west across the Atlantic. By August 12 it was assaulting Puerto Rico and still growing in intensity.

Weather forecasters watched Camille closely and tried to predict its path, which was erratic. They sent out aircraft to locate and study the hurricane, hoping to warn people in its path as well as to obtain scientific knowledge of the storm. As Camille passed near the western tip of Cuba, timely warnings by forecasters saved many people in that country from injury or even death. Camille then headed for the United States—but where would it strike?

*The spiral shape of a hurricane, as seen from a satellite.*

Hurricane experts worked furiously to answer that question. They collected every bit of information they could from planes, satellites, and ground observations. Then they fed this information into computers. Their computer models predicted that Camille would strike the southern part of Florida. But computer forecasting was new in 1969, and the forecasters' own experience with similar storms made them treat this prediction with caution. Many of them felt that the hurricane would strike the northwest coast of Florida instead. On the morning of August 16, the National Hurricane Center in Miami issued a conservative warning based on these conclusions. The warning stretched from St. Marks, Florida, to Biloxi, Mississippi. The next day, scientists observed erratic changes in the behavior of Camille, causing them to extend their warning westward as far as New Orleans, Louisiana.

*Offshore wind on waves.*

During this time, Camille had been sitting off the coast, growing in strength. By the morning of August 17, it became clear that Camille probably would avoid Florida altogether and instead would assault the coast somewhere between Louisiana and Mississippi. The Hurricane Center issued strong warnings, and many people wisely evacuated the coastal areas, heading inland to safety. Those who ignored the warnings, or who waited until it was too late, regretted their decisions. Some paid for this mistake with their lives.

*Hurricane Allen at Corpus Christi, Texas, 1980.*

The storm charged the Mississippi coast near the city of Gulfport. With winds up to 190 miles (306 km) per hour, Camille knocked down telephone poles, power lines, and nearly everything else in its path. Residents of the storm-ravaged communities feared for their lives. Structures near the coast were the hardest hit. The buildings not destroyed by the wind were deluged by a tidal surge 25 feet (7.6 m) high. Many of the people in these buildings were killed.

When the storm was finally over, the death toll was 330, and the damage to property was estimated at more than one and a half billion dollars. While the forecasters' initial predictions had been inaccurate, their final warnings still allowed hundreds, perhaps thousands, of people to seek refuge in time to save their lives. Who knows how much greater the death toll would have been without the efforts of these scientists?

*The violent energy of a hurricane.*

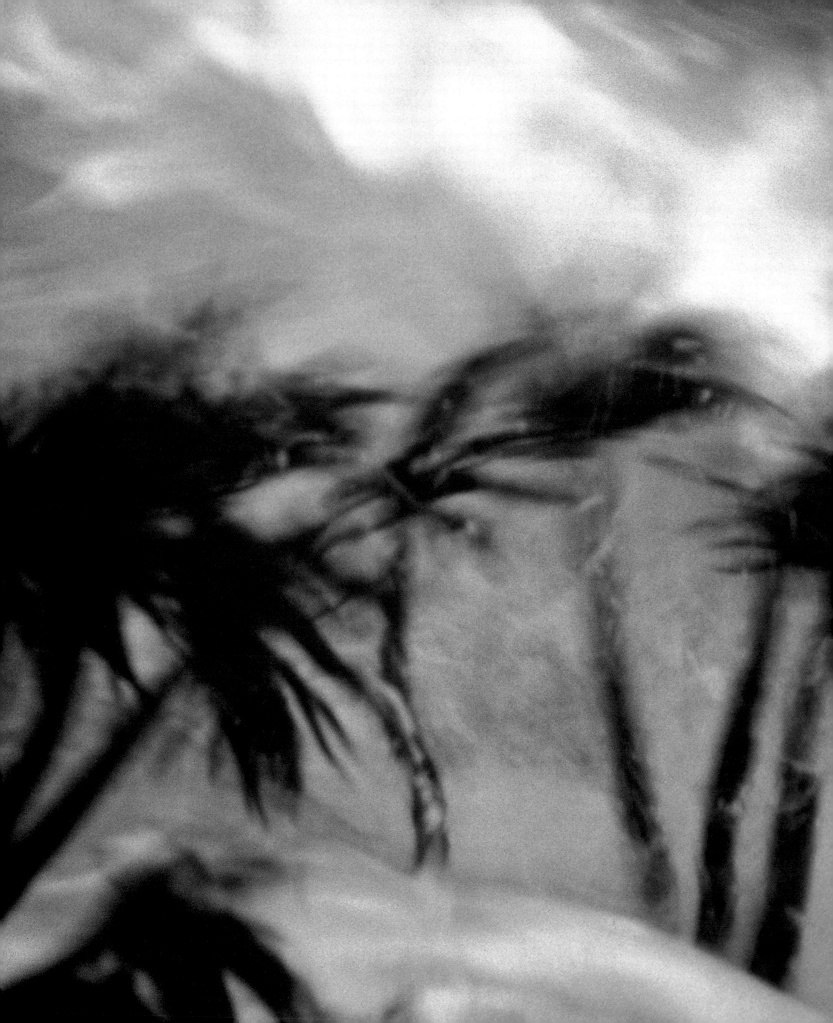

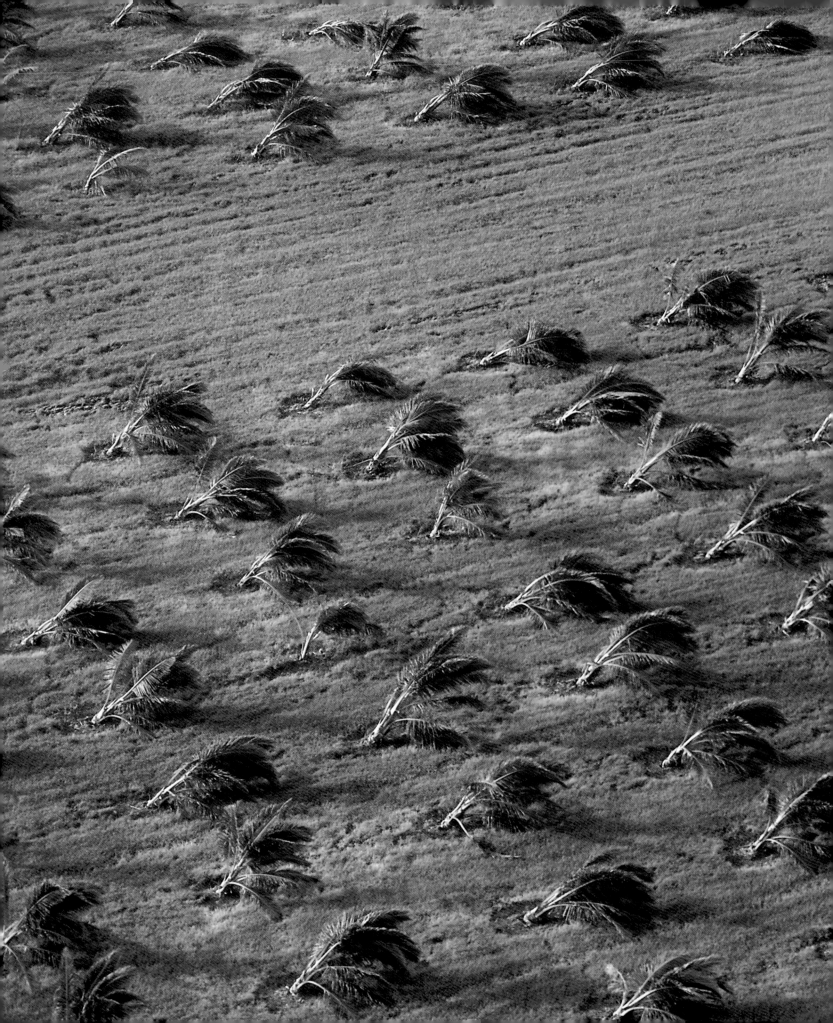

Camille was one of the most powerful hurricanes ever to hit the United States. Yet the death toll from some other tropical cyclones, both in North America and in many other parts of the world, has been much, much greater. In 1737, a storm whipped across the southern coast of India, bringing waves and tidal surges that drowned an estimated 300,000 people. Even in the United States, the devastation from other storms has been much worse than that caused by Camille. In 1900, for example, a violent hurricane deluged the Texas city of Galveston, killing 6,000 people.

It is not necessarily the intensity of the storm that determines its deadliness. Rather, it is the vulnerability of the people in its path. This is why the predictions of hurricane forecasters are so important. With adequate warning, people can prepare to ride out the storm or flee it, resulting in great savings of both lives and property.

*A Florida tree farm suffers the onslaught of Hurricane Andrew, 1992.*

Scientists studying the weather today have better tools at their disposal than did the experts who tracked Camille. Modern super-computers, sophisticated satellites, and more sensitive instruments have all improved the accuracy of weather forecasting. But scientists face serious challenges. Because the population is increasing in vulnerable areas, many hurricanes still cause terrible devastation. This makes the science of hurricane prediction even more important. As it continues to improve, many more thousands of lives around the world can be saved by timely and accurate warnings.

*1989: Sailboats in the grip of Hurricane Hugo.*

It is unlikely that we will ever be able to control the awesome forces that combine to produce hurricanes. Like earthquakes and volcanoes, *Hurricanes* represent the untamable power of nature. Advances in science and technology may save us from them, but only by helping us to get out of their way.